Peppercorns

Growing Practices and Nutritional Information

Roby Jose Ciju

CONTENTS

Introduction 7
   Black Peppercorns 7
   White Peppercorns 7
   Green Peppercorns 8
   Red Peppercorns 8
Origin and Distribution 8
Taxonomy 8
Botanical description 9
Common Names 9
Varieties 10
Nutrition and Health Benefits of Peppercorns 10
   Nutrition in Blackpepper in 100 grams of edible portion 11
   Nutrition in Whole Blackpepper 13
   Nutrition in Ground Blackpepper 15
   Nutrition in White pepper 17
   Nutrition in Ground White Pepper 18
Pungency in Peppercorns 20
Aroma in Peppercorns 20
Uses of Peppercorns 20
Growing Practices for Black Pepper 21
   Climate Requirements 21
   Soil Requirements 21
   Propagation 21
   Planting Method 21
   Planting time 22
   Field Preparation 22
   Pruning of Support Plants 22
   Irrigation 22
   Fertilizer Application and Manuring 23
   Intercultural Operations 23
   Harvesting 23
   Curing of Black Pepper 23
   Yield 24
   Economic Life of a Black Pepper Plantation 24

Grading of Black Pepper...24

Grading of White peppercorns....................................25

Storage...25

Peppercorns: Production Practices and Nutrition Information

Introduction

Peppercorns are one of the most used Indian spices across the globe. Peppercorns are produced from pepper plants. Scientific name of pepper is *Piper nigrum* and this plant belongs to the family Piperaceae. It is an evergreen, tender, perennial climbing vine that prefers partial shade for its growth. The plant reaches up to a height of 400cm and spread of 400cm upon full growth.

 Dried fruits of pepper plants are known as "peppercorns". There are black, white, green and red peppercorns depending on the harvesting stage and processing of the mature fruits. Most popular peppercorns are black peppercorns and white peppercorns. Black pepper contains about 3% essential oil, and white pepper contains 1.5%. In addition to these, dried green peppercorns and red peppercorns are also available in the market.

Black Peppercorns

Black pepper is dried fruits of pepper plants. Mature green fruits are harvested and dried to get black peppercorns. Whole green pepper spikes are harvested when the fruits reach maturity and first few berries are turning red in colour. Harvested pepper spikes are stored and individual berries are separated by trampling on a clean floor. Pepper berries are taken to sun drying. Under sun drying, green peppercorns slowly turn into dark black in colour.

White Peppercorns

Fully ripened pepper berries are used for making white pepper. The outer hull of the ripened berries is removed by soaking the berries in slow-running water for about one week. After one week outer hull disintegrates and kernels are obtained. Kernels or seed grain is then dried and sold as white pepper. White pepper is expensive than black pepper. White pepper has same pungency as that of black pepper but its flavour remains altered due to the removal of outer hull and partial loss of some aroma compounds.

Green Peppercorns

Green pepper is early-harvested immature green berries of pepper plants. Green pepper corns are made by pickling freshly harvested unripe fruits of pepper in salt or vinegar or by quick drying at elevated temperature or in vacuum. Since green pepper corns are unripe it has a green flavour and is less pungent. Quick dried green peppercorns are highly aromatic but at the same time less pungent. Hence it is used for many delicate food preparations to add aroma to it. Green pepper corns are used in preparing pepper steaks, sauces that accompany non-vegetarian food items, and as a garnish to cold foods.

Red Peppercorns

Red pepper corns are made from ripened red pepper corns by pickling them while ensuring that their red colour is retained. This is a very rare commodity.

Origin and Distribution

Black pepper is believed to be originated in the west coast of southern India. Pepper is cultivated in South East Asian countries such as Malaysia, Madagascar, Indonesia, Thailand, Vietnam, China, and Srilanka and also in Brazil.

Taxonomy

Kingdom	Plantae – Plants
Subkingdom	Tracheobionta – Vascular plants
Superdivision	Spermatophyta – Seed plants
Division	Magnoliophyta – Flowering plants
Class	Magnoliopsida – Dicotyledons
Subclass	Magnoliidae
Order	Piperales
Family	Piperaceae – Pepper family
Genus	Piper L. – pepper
Species	Piper nigrum L. – black pepper

Source: USDA Plant database

Botanical description

Plant	Pepper plant is a dicot, perennial plant with woody climbing vines which grows up to 4 metres in height on supporting trees, or poles
Stem	Stem is swollen a the nodes; climbing roots are produced from the nodes and these roots help the plants to climb on the support; that is, stem is a climbing vine; three types of vines are seen: terminal stem, stolons or runners and lateral branches
Leaves	Stems produce alternate leaves which are shiny, dark green and ovate in shape with a size of 8-20 x 4-12 cm. Branches are produced from the dorsal buds
Inflorescence	Inflorescence develops on current year's flush which is opposite to a leaf
Flower	Small whitish flowers are borne on hanging spikes; mature spike varies in length from 10-30 cm in length, supporting up to 100-150 small sessile flowers without perianth. The period from flowering to harvest is about 180-240 days
Fruit	Fruit is a small, sessile indehiscent berry of 4-6 mm in diameter which is green in colour while immature and turns yellowish to reddish as it ripens

Common Names

Black pepper, White pepper, Pepper, Madagascar pepper, Poivre noir, Pimienta negra, Filfil aswad, Hu-chiao, Peper, Poivre, Pfeffer, Pepe, Kosho, Pimenta, Pyerets, Pimienta, Peppar, Kali mirch, Kala morich, Golmorich, Kalamari, Kalomirich, Kare menasu, Marutis, Kurumalaku, nallamulaku, Mira, Gol-maricha, Marich ushana, Hapusha, Milagu, Miriyala tige, Siah mirch.

Varieties

Panniyur 1 is a high yielding hybrid variety of India. Other two commercial varieties known to spice trading markets are 'Malabar Aleppey' and 'Malabar Tellicherry'.

Nutrition and Health Benefits of Peppercorns

Black pepper has antibacterial properties and hence used to preserve food items. Black pepper helps the body to fight against infections due to its antibacterial properties. It has also anti-inflammatory properties. Black pepper aids in proper digestion and is therefore good for stomach. It promotes sweating and is also a diuretic (promotes urination). Black pepper is also a carminative (expels gas out of the body in a healthy way and inhibits gas formation within the body).

Black pepper corns are used for weight loss also as the outer layer of peppercorns helps in breaking down of fat cells. In traditional medicines, black pepper is used as an ingredient for cough and cold medicines due to its expectorant property (breaks down phlegm deposits in the respiratory tract and helps to expel it through coughing and thus eliminates the phlegm from the body).

Black pepper has high antioxidant property also. Antioxidants prevent cell damage caused by free radicals and thus prevents occurrence of cancer and age-related diseases. Piperine, major active principle found in pepper promotes neurological health. Adding black pepper in food items not only increases its taste but it also enhances the availability of other nutrients present in the food to the body system.

Black pepper is rich source of calcium (443mg/100g), potassium (1329mg/100g), and dietary fiber (25.3 g/100g). It also contains good amounts of iron, magnesium and phosphorous. Black pepper also contains vitamin A (547 IU/100g), vitamin K (163.7µg/100g), and vitamin E (1.04mg/100g).

Nutrition in Blackpepper in 100 grams of edible portion

Nutrient	Unit	Value per 100 g
Proximates		
Water	g	12.46
Energy	kcal	251
Protein	g	10.39
Total lipid (fat)	g	3.26
Carbohydrate, by difference	g	63.95
Fiber, total dietary	g	25.3
Sugars, total	g	0.64
Minerals		
Calcium, Ca	mg	443
Iron, Fe	mg	9.71
Magnesium, Mg	mg	171
Phosphorus, P	mg	158
Potassium, K	mg	1329
Sodium, Na	mg	20
Zinc, Zn	mg	1.19
Vitamins		
Vitamin C, total ascorbic acid	mg	0
Thiamin	mg	0.108
Riboflavin	mg	0.18
Niacin	mg	1.143
Vitamin B-6	mg	0.291
Folate, DFE	µg	17
Vitamin B-12	µg	0
Vitamin A, RAE	µg	27
Vitamin A, IU	IU	547
Vitamin E (alpha-tocopherol)	mg	1.04

Vitamin D (D2 + D3)	µg	0
Vitamin D	IU	0
Vitamin K (phylloquinone)	µg	163.7
Lipids		
Fatty acids, total saturated	g	1.392
Fatty acids, total monounsaturated	g	0.739
Fatty acids, total polyunsaturated	g	0.998
Fatty acids, total trans	g	0
Cholesterol	mg	0
Other		
Caffeine	mg	0

Source: USDA Nutrient Database

Nutrition in Whole Blackpepper

Nutrient	Unit	1 tsp, whole = 2.9g	1 dash = 0.1g
Proximates			
Water	g	0.36	0.01
Energy	kcal	7	0
Protein	g	0.3	0.01
Total lipid (fat)	g	0.09	0
Carbohydrate, by difference	g	1.85	0.06
Fiber, total dietary	g	0.7	0
Sugars, total	g	0.02	0
Minerals			
Calcium, Ca	mg	13	0
Iron, Fe	mg	0.28	0.01
Magnesium, Mg	mg	5	0
Phosphorus, P	mg	5	0
Potassium, K	mg	39	1
Sodium, Na	mg	1	0
Zinc, Zn	mg	0.03	0
Vitamins			
Vitamin C, total ascorbic acid	mg	0	0
Thiamin	mg	0.003	0
Riboflavin	mg	0.005	0
Niacin	mg	0.033	0.001
Vitamin B-6	mg	0.008	0
Folate, DFE	µg	0	0
Vitamin B-12	µg	0	0
Vitamin A, RAE	µg	1	0
Vitamin A, IU	IU	16	1
Vitamin E (alpha-tocopherol)	mg	0.03	0
Vitamin D (D2 + D3)	µg	0	0

Vitamin D	IU	0	0
Vitamin K (phylloquinone)	µg	4.7	0.2
Lipids			
Fatty acids, total saturated	g	0.04	0.001
Fatty acids, total monounsaturated	g	0.021	0.001
Fatty acids, total polyunsaturated	g	0.029	0.001
Fatty acids, total trans	g	0	0
Cholesterol	mg	0	0
Other			
Caffeine	mg	0	0

Source: USDA Nutrient Database

Nutrition in Ground Blackpepper

Nutrient	Unit	1 tsp, ground = 2.3g	1 tbsp, ground = 6.9g
Proximates			
Water	g	0.29	0.86
Energy	kcal	6	17
Protein	g	0.24	0.72
Total lipid (fat)	g	0.07	0.22
Carbohydrate, by difference	g	1.47	4.41
Fiber, total dietary	g	0.6	1.7
Sugars, total	g	0.01	0.04
Minerals			
Calcium, Ca	mg	10	31
Iron, Fe	mg	0.22	0.67
Magnesium, Mg	mg	4	12
Phosphorus, P	mg	4	11
Potassium, K	mg	31	92
Sodium, Na	mg	0	1
Zinc, Zn	mg	0.03	0.08
Vitamins			
Vitamin C, total ascorbic acid	mg	0	0
Thiamin	mg	0.002	0.007
Riboflavin	mg	0.004	0.012
Niacin	mg	0.026	0.079
Vitamin B-6	mg	0.007	0.02
Folate, DFE	µg	0	1
Vitamin B-12	µg	0	0
Vitamin A, RAE	µg	1	2
Vitamin A, IU	IU	13	38
Vitamin E (alpha-tocopherol)	mg	0.02	0.07
Vitamin D (D2 + D3)	µg	0	0

Vitamin D	IU	0	0
Vitamin K (phylloquinone)	µg	3.8	11.3
Lipids			
Fatty acids, total saturated	g	0.032	0.096
Fatty acids, total monounsaturated	g	0.017	0.051
Fatty acids, total polyunsaturated	g	0.023	0.069
Fatty acids, total trans	g	0	0
Cholesterol	mg	0	0
Other			
Caffeine	mg	0	0

Source: USDA Nutrient Database

Nutrition in White pepper

Nutrition in 100 grams edible portion of white pepper

Nutrient	Unit	Value per 100 g
Proximates		
Water	g	11.42
Energy	kcal	296
Protein	g	10.4
Total lipid (fat)	g	2.12
Carbohydrate, by difference	g	68.61
Fiber, total dietary	g	26.2
Minerals		
Calcium, Ca	mg	265
Iron, Fe	mg	14.31
Magnesium, Mg	mg	90
Phosphorus, P	mg	176
Potassium, K	mg	73
Sodium, Na	mg	5
Zinc, Zn	mg	1.13
Vitamins		
Vitamin C, total ascorbic acid	mg	21
Thiamin	mg	0.022
Riboflavin	mg	0.126
Niacin	mg	0.212
Vitamin B-6	mg	0.1
Folate, DFE	µg	10
Vitamin B-12	µg	0
Vitamin A, RAE	µg	0
Vitamin A, IU	IU	0
Vitamin D (D2 + D3)	µg	0
Vitamin D	IU	0
Lipids		
Fatty acids, total saturated	g	0.626
Fatty acids, total monounsaturated	g	0.789

Fatty acids, total polyunsaturated	g	0.616
Cholesterol	mg	0

Source: USDA Nutrient Database

Nutrition in Ground White Pepper

Nutrient	Unit	1 tsp, ground = 2.4g	1 tbsp, ground = 7.1g
Proximates			
Water	g	0.27	0.81
Energy	kcal	7	21
Protein	g	0.25	0.74
Total lipid (fat)	g	0.05	0.15
Carbohydrate, by difference	g	1.65	4.87
Fiber, total dietary	g	0.6	1.9
Minerals			
Calcium, Ca	mg	6	19
Iron, Fe	mg	0.34	1.02
Magnesium, Mg	mg	2	6
Phosphorus, P	mg	4	12
Potassium, K	mg	2	5
Sodium, Na	mg	0	0
Zinc, Zn	mg	0.03	0.08
Vitamins			
Vitamin C, total ascorbic acid	mg	0.5	1.5
Thiamin	mg	0.001	0.002
Riboflavin	mg	0.003	0.009
Niacin	mg	0.005	0.015
Vitamin B-6	mg	0.002	0.007
Folate, DFE	µg	0	1
Vitamin B-12	µg	0	0
Vitamin A, RAE	µg	0	0
Vitamin A, IU	IU	0	0
Vitamin D (D2 + D3)	µg	0	0

Vitamin D	IU	0	0
Lipids			
Fatty acids, total saturated	g	0.015	0.044
Fatty acids, total monounsaturated	g	0.019	0.056
Fatty acids, total polyunsaturated	g	0.015	0.044
Cholesterol	mg	0	0

Source: USDA Nutrient Database

Pungency in Peppercorns

Among four types of peppercorns, pungency is strongest in white pepper corns, then in black pepper corns. Pungency is weakest in green pepper corns. The pungent principle in peppercorns is an alkaloid-analog compound, called piperine.

Aroma in Peppercorns

Blackpepper corns are highly aromatic followed by green pepper corns. Green peppercorns have an immature herbaceous aroma. Red peppercorns are sweet in taste with a typical pepper aroma.

Uses of Peppercorns

Pickled green peppercorns: Green pepper corns pickled in brine or vinegar is available in the market. Pickled red pepper corns are also popular among consumers. Black peppercorns and white peppercorns are not usually pickled.

Black peppercorns are used as a spice and food seasoning agent. Black pepper is used for preparing spicy food preparations, particularly non-vegetarian food items. Black pepper is an ingredient of garam masala, South Indian sambar masala, and the Anglo–Indian curry powder. Black pepper powder is used as a spice for black tea which has medicinal benefits. Ground black pepper is used in traditional medicines.

Dried peppercorns are used for extracting pepper oil which has numerous applications in medicinal and cosmetic industries.

Growing Practices for Black Pepper

Climate Requirements

Black pepper is a plant of humid tropics and flourishes in warm, moist climate. An annual rainfall of 2500 mm is suitable for its cultivation. Black pepper can tolerate a temperature range of 10-40 degree Celsius. It grows well from the mean sea level up to an altitude of 1200 meters.

Soil Requirements

Black pepper grows well in well-drained clay-loam soil which is rich in organic matter (humus).

Propagation

Black pepper is propagated from cuttings of the runner shoots which originate from the base of the vines. Cuttings from the lateral fruiting branches may also be used for propagation. 40 to 50 centimetres long cuttings are generally used.

Planting Method

In tropics, vines from which cuttings are made for propagation are separated from the mother plant during February-March. Vines are cut into equally long cuttings and each cutting needs to have at least 2 -3 nodes. Cuttings are then planted in polybags or bamboo baskets or well-prepared nursery beds.

Raising of cuttings in polybags or bamboo baskets

Containers are filled with well-prepared growing media and cuttings are planted in them @one cutting per polybag or bamboo basket. Cutting is planted with at least one node below the soil level. After planting, polybags or bamboo baskets are kept in shade.

Raising cuttings in nursery beds

Nursery beds are prepared under shade by ploughing and levelling the soil followed by bed making. Then nursery beds are filled with fine soil and a suitable growing media before planting the cuttings in beds at appropriate spacing. Each cutting is planted with at least one node below the soil.

Planting time

Cuttings will be ready for transplanting during June-July. In India, black pepper is planted at the onset of south-west monsoon which begins in the month of June.

Live Support Plants or Standard Plants Used for Black Pepper Plantations

In India, most commonly used support plants for training black pepper climbing vines are *Erythrina indica*, *Garuga pinnata*, *Spondias mangifera*, and silver oak (*Grevillea robusta*).

In some countries dead hardwood supports are used as support for black pepper plants.

Field Preparation

Main field or planting site is prepared well in advance by ploughing and levelling of the land. Pits of 0.5m x 0.5m x 0.5 m size are prepared in the site at a spacing of 3-4 meters apart in each direction. With the fist showers, cuttings of the selected support plant are planted in these pits as standards. After one month, cuttings of black pepper are planted @ 2-3 cuttings/standard about 30 centimeters away from the base. Each cutting is secured to the support by using a fiber.

Pepper may also be grown in association with mango, coconut and jackfruit trees.

Pruning of Support Plants

Pruning of support plants or standards may be necessary just before flowering and fruiting stages of plant production. Too much shade during flowering and fruiting stages may adversely affect the plant production. In tropics, pruning of standards is done during March – April.

Irrigation

In humid tropics, black pepper is mainly grown as a rainfed crop.

Fertilizer Application and Manuring

Soon after planting, each vine may be supplied with about 10 kilograms of farm yard manure or compost as a basal dose. The manure or compost needs to be applied around the base of the vine at a distance of 30 cm from the base and up to a depth of about 15 cm. After applying the manures in the soil, it needs to be incorporated into the top soil by using a fork. One kilogram of lime may be applied in alternate years during April-May in order to improve soil fertility.

On an average each vine requires about 100 grams of nitrogen, 160 grams of phosphorous and 60 grams of potassium per year in two equal doses when they are 3-4 years old.

Intercultural Operations

Digging around the standards and base of the vines up to a diameter of 2 meters is recommended at least twice a year. First digging is done during August-September and second digging is done during October-November.

If the black pepper plantation is located in hill slopes, adequate soil conservation measures such as cover cropping need to be adopted.

Harvesting

Black pepper starts bearing fruits from third year of planting. Flowering begins in June and it takes about 6-8 months from flowering to harvesting.

The season for harvesting mature berries in India is November – February in plains and January – March in hills.

Harvesting is done by plucking the spikes when one or two berries turn colour from green to yellow to red.

Curing of Black Pepper

Freshly harvested spikes are spread on a clean floor or on clean mats. Berries are separated from spikes by trampling. Separated berries are cleaned and dried in the sun for up to one week until the outer skin shrink and turn to black colour. Thus black pepper of commercial use is obtained. Recovery of black pepper is about 33 percent of ripe green berries.

White pepper is prepared from fully ripened berries. All red coloured fully ripened berries are taken and their outer skin along with pulp is removed before drying them.

Recovery of white pepper is 25 percent of the ripe red berries.

Yield

Even though the black pepper plant starts bearing from third year onwards, yield becomes stable from 7 th or 8 th year onwards. Generally speaking, one hectare of black pepper plantation yields about 600-800 Kg black pepper.

Optimum yield of a black pepper plantation at low-input farming system is,

Unprocessed green pepper	6 tons/ha
Sundried black pepper	2 tons/ha
Washed and dried white pepper	1.67 tons/ha

In plantations, with higher inputs, yields may be 8-9 tons/ha of green pepper in the first harvest and 12-20 tons/ha in the sixth or seventh harvest.

Economic Life of a Black Pepper Plantation

A black pepper plantation begins to bear in 3-4 years, reach full production after 7 years and with an economical life over 20 years.

Under good cultural management practices, a black pepper plantation remains productive up to 25 years. Under best practices, the vine may live up to 40-60 years.

Grading of Black Pepper

Grading of black pepper is normally done based on their source of origin.

Source of Origin	Description
Indian Black Pepper	Indian black pepper grades are: Malabar and Tellicherry. Both grades are very aromatic and pungent.
Malaysian Black	Malaysian blackpepper grade is Sarawak

24

Pepper	black pepper. Sarawak pepper is mild, less aromatic and fruity. It is a small-fruited black pepper that takes on a greyish colour during storage. Sarawak pepper is native to the Malaysian region of Borneo.
Indonesian Black Pepper	Indonesian black pepper grade is Lampong. Lampung hails from island of Sumatra. It is a small-fruited black pepper that takes on a greyish colour during storage; Lampong pepper is pretty hot with a less-developed aroma.

Grading of White peppercorns

Indonesian white peppercorns	White Muntok and Sarawak pepper comes from Indonesia. Sarawak is light-coloured; the best quality white peppercorn is known as Sarawak Cream Label.
Brazil white peppercorns	Brazil white pepper, i.e. Belém, named after its main port, has mild quality, and a poorer flavour. It is less reputed in the international trade.

In addition to this, Vietnam pepper (blackpepper and whitepepper) and Kampot Pepper (black, green, red, and white) of Cambodia are also available in the international market.

Storage

Since peppercorns lose its typical aroma and flavour through evaporation, it is advised to store peppercorns in air tight containers. It also helps to preserve its spiciness for a longer period.